月　球

我们的太空邻居

U0240965

THE MOON

Our Neighbor in Space

（英国）埃伦·劳伦斯／著　　刘　颖／译

江苏凤凰美术出版社

著作权合同登记图字：10-2022-144

图书在版编目（CIP）数据

月球：我们的太空邻居 / （英）埃伦·劳伦斯著；

刘颖译 . -- 南京：江苏凤凰美术出版社，2025.1.

（环游太空）. -- ISBN 978-7-5741-2027-3

Ⅰ . P184-49

中国国家版本馆 CIP 数据核字第 2024ZJ7773 号

策　　　　划　朱　婧
责 任 编 辑　高　静　奚　鑫
责 任 校 对　王　璇
责任设计编辑　樊旭颖
责 任 监 印　生　媛
英 文 朗 读　C.A.Scully
项 目 协 助　邵楚楚　乔一文雯

丛　书　名　环游太空
书　　　名　月球：我们的太空邻居
著　　　者　（英国）埃伦·劳伦斯
译　　　者　刘　颖
出 版 发 行　江苏凤凰美术出版社（南京市湖南路 1 号 邮编：210009）
印　　　刷　南京新世纪联盟印务有限公司
开　　　本　710 mm×1000 mm　1/16
总 印 张　18
版　　　次　2025 年 1 月第 1 版
印　　　次　2025 年 1 月第 1 次印刷
标 准 书 号　ISBN 978-7-5741-2027-3
总 定 价　198.00 元（全 12 册）

版权所有　侵权必究
营销部电话：025-68155675　营销部地址：南京市湖南路 1 号
江苏凤凰美术出版社图书凡印装错误可向承印厂调换

目录 Contents

书中加粗的词语见词汇表解释。

Words shown in **bold** in the text are explained in the glossary.

欢迎来到月球
Welcome to the Moon

想象一个距离地球数十万千米的世界。

Imagine a world that is hundreds of thousands of kilometers from Earth.

那里的地面布满了岩石和厚厚的灰色尘埃。

The ground is covered with rocks and thick, gray dust.

那里还有高山、平原和巨大的陨石坑。

This place is home to tall mountains, flat **plains**, and gigantic holes called **craters**.

白天极为炽热。

During the day, it is scorching hot.

夜晚，它比地球上最冷的地方还要冷得多。

At night, it is much colder than the coldest place on Earth.

欢迎来到月球！

Welcome to the Moon!

陨石坑 Crater

这张照片里，一位名叫查尔斯·杜克的宇航员正站在月球上。除了地球之外，月球是人类在太空中唯一踏足过的地方。

This photo shows an **astronaut** named Charles Duke standing on the Moon. Other than Earth, the Moon is the only place in space where humans have ever stood.

宇航员 Astronaut

太空邻居
Space Neighbors

月球与我们的地球家园一起在太空中运行。

The Moon and our home planet Earth are traveling through space together.

当它们在太空中移动时，月球同时环绕着地球公转。

As they move through space, the Moon is **orbiting**, or circling, Earth.

月球绕地球公转一圈需略多于27天。

It takes the Moon just over 27 days to orbit Earth once.

当我们在夜空中看到月亮时，它呈现白色或灰白色。

When we see the Moon in the night sky, it looks white or grayish-white.

那是因为太阳光照在月球上，月球将光线反射到人们眼中。

That's because the Sun is shining on the Moon and lighting it up.

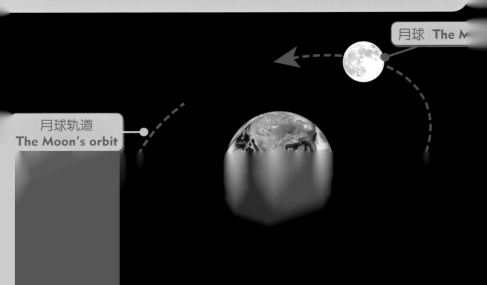

月球 The M

月球轨道
The Moon's orbit

从地球上看月球，它似乎总在改变形状。但事实并非如此。当月球围绕地球公转时，我们可以看到月球被照亮的不同部分。

From Earth, it looks as if the Moon is changing shape. It doesn't really, though. As it travels around Earth, we see different parts of its bright, shining surface.

这些照片展示了我们从地球上看到的形状各异的月亮。

These photos show the different ways that we see the Moon from Earth.

7

太阳系 The Solar System

地球和月球围绕着太阳做一个巨大的圆周运动。

Earth and the Moon are traveling in a big circle around the Sun.

地球是围绕太阳公转的八大行星之一。

Earth is one of eight **planets** circling the Sun.

八大行星分别是水星、金星、地球、火星、木星、土星、天王星和海王星。

The planets are called Mercury, Venus, Earth, Mars, Jupiter, Saturn, Uranus, and Neptune.

有些行星也有卫星作为它们的"邻居"。

Some of the other planets have **moons** as neighbors, too.

除了行星和卫星，由岩石构成的小行星和冰冻的彗星也围绕着太阳公转。

In addition to planets and moons, rocky **asteroids** and icy **comets** orbit the Sun.

太阳和它周围的天体共同组成了"太阳系"。

Together, the Sun and its family of space objects are called the solar system.

大多数围绕太阳的小行星都集中在被称为"小行星带"的环状带中。

Most of the asteroids orbiting the Sun are in a ring called the asteroid belt.

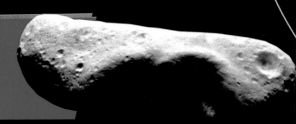

小行星 An asteroid

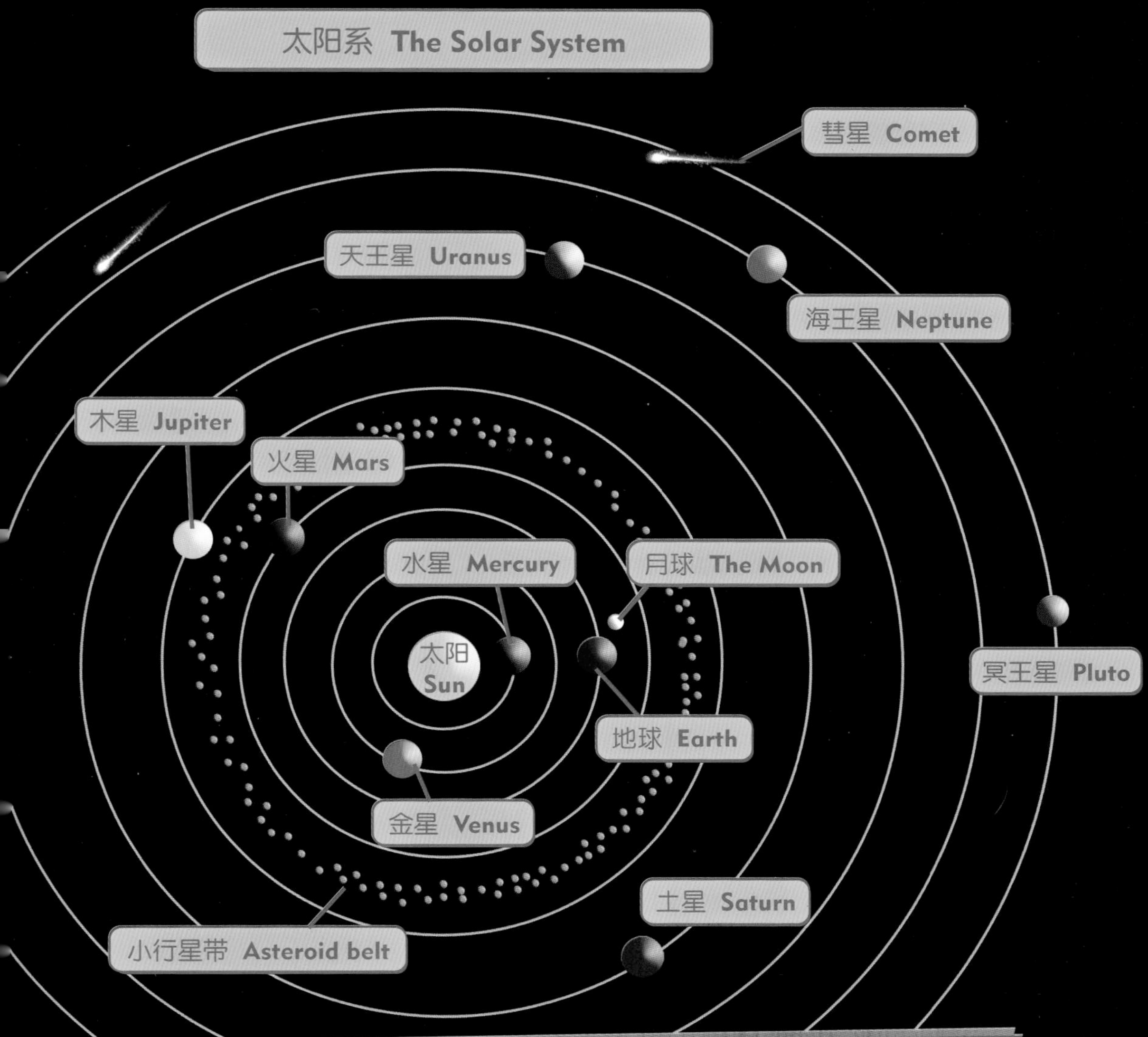

太阳系 **The Solar System**

彗星 **Comet**

天王星 **Uranus**

海王星 **Neptune**

木星 **Jupiter**

火星 **Mars**

水星 **Mercury**

月球 **The Moon**

太阳
Sun

地球 **Earth**

金星 **Venus**

冥王星 **Pluto**

土星 **Saturn**

小行星带 **Asteroid belt**

太阳系里还有更小的星球，它们被称为"矮行星"。冥王星就是一颗矮行星。

The solar system is home to small planets, called **dwarf planets**. Pluto is a dwarf planet.

近距离观察月球
A Closer Look at the Moon

月球的直径仅3 000多千米。

The Moon is just over 3,000 kilometers wide.

它主要由岩石构成，中心是个铁球。

It is made mostly of rock with a ball of iron in the center.

地球覆盖着一层厚厚的气体，被称为"大气层"。

Earth is covered with a thick layer of **gases** called an **atmosphere**.

正是这层气体使地球的天空看起来是蓝色的。

It's these gases that make Earth's sky look blue.

月球没有大气层。

The Moon does not have an atmosphere.

所以，无论是白天还是黑夜，月球的天空看起来都是黑色的。

So from the Moon the sky looks black whether it is day or night.

月球有多大?
How Big Is the Moon?

地球 Earth

月球 The Moon

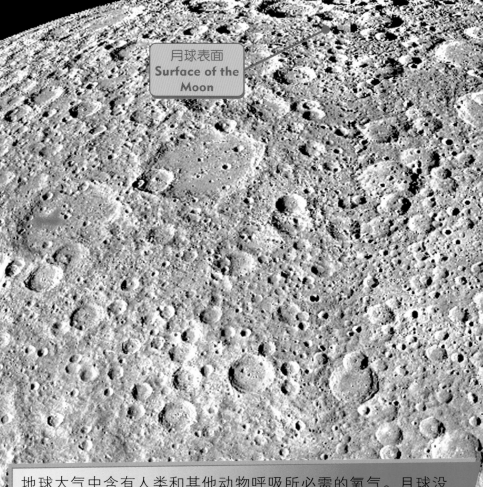

黑色的天空 Black sky

月球表面
**Surface of the
Moon**

地球大气中含有人类和其他动物呼吸所必需的氧气。月球没有大气层，因此月球上没有可供呼吸的氧气。

Earth's atmosphere contains **oxygen** that humans and other animals need to breathe. Because the Moon has no atmosphere, there is no oxygen to breathe on the Moon.

仰天望月
The Moon Up Close

当你抬头望向月亮时，你能看到它的表面有大片暗色的区域。

If you look up at the Moon, you can see large dark areas on its surface.

数百年前，人们猜测这些暗色区域可能是海洋。

Hundreds of years ago, people thought these dark areas might be oceans.

今天，我们知道那些区域是宽阔平坦的平原，由光滑的深色岩石构成。

Today, we know they are wide, flat plains made of dark, smooth rock.

通过望远镜也能看到月球上的山脉。

It's also possible to see mountains on the Moon through a telescope.

其中一些高达5千米。

Some of the Moon's mountains are up to 5 kilometers tall.

月球表面覆盖着一层厚厚的尘埃和岩石。有些岩石有卡车那么大。

The surface of the Moon is covered with a thick layer of dust and rocks. Some of the rocks are as big as trucks.

平坦的平原
Smooth plains

山脉 **Mountains**

陨石坑 **Craters**

13

巨大的陨石坑
A Gigantic Crater

月球表面覆盖着数十万个陨石坑。

The surface of the Moon is covered with hundreds of thousands of craters.

这些陨石坑是由小行星和彗星等天体撞击月球而形成的。

These craters are made by objects such as asteroids and comets that hit the Moon.

许多陨石坑直径达数百千米。

Many of the craters are hundreds of kilometers wide.

最大的那些陨石坑被称为"撞击盆地"。

The biggest craters are known as **impact basins**.

月球上最大的撞击盆地直径为2 600千米。

The largest impact basin on the Moon is 2,600 kilometers across.

如果这个巨大的陨石坑在地球上,它能覆盖美国面积的一半!

If this gigantic crater was on Earth, it could cover half of the United States!

月球上有个巨大的撞击盆地,名叫南极–艾特肯盆地。它位于月球南极。盆地深度超过13千米。

The Moon's gigantic impact basin is called the South Pole-Aitken basin. It is at the Moon's south pole. The basin is over 13 kilometers deep.

南极－艾特肯盆地
South Pole-Aitken Basin

高山 tall mountains

较小的
山脉 smaller mountains
and hills

中等高度
的区域 medium-height areas

极低的区域 very low areas

这张月球图片是计算机生成的。不同颜色代表了月球表面不同区域的高度。
This picture of the Moon was created on a computer. The colors show the different heights of the land.

月球漫步
Walking on the Moon

1969年7月，3名宇航员乘坐"哥伦比亚号"太空飞船离开地球。

当他们抵达月球后，宇航员迈克尔·柯林斯留守在"哥伦比亚号"上。

尼尔·阿姆斯特朗和巴兹·奥尔德林乘"鹰号"登月舱在月球表面着陆。

1969年7月20日，阿姆斯特朗和奥尔德林成为登上月球的第一批人！

这两名宇航员在月球上探索了2小时36分钟。

之后，他们驾驶"鹰号"返回"哥伦比亚号"，再回到地球。

In July 1969, three astronauts left Earth aboard a spacecraft named *Columbia*.

When they reached the Moon, astronaut Michael Collins stayed aboard *Columbia*.

Neil Armstrong and Buzz Aldrin flew down to the Moon's surface in a spacecraft called *the Eagle*.

On July 20, 1969, Armstrong and Aldrin became the first people to walk on the Moon!

The astronauts explored the Moon for two hours and 36 minutes.

Then they flew *the Eagle* back to *Columbia*, and headed home.

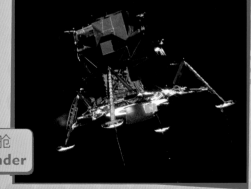

"鹰号"登月舱
The Eagle Lander

在月球漫步期间，宇航员收集了月球上的岩石。他们还使用电视摄像机拍下了月球的照片和录像。

During their moonwalk, the astronauts collected moon rocks. They also took photos and filmed the Moon with TV cameras.

巴兹·奥尔德林
Buzz Aldrin

宇航员的头套里有氧气，可供他们呼吸。宇航服还保护他们免受月球上的极端高温和低温影响。

Inside the astronauts' helmets there was oxygen so they could breathe. Their spacesuits protected them from the extreme hot and cold temperatures on the Moon.

研究月球
Studying the Moon

自首次月球漫步以来，有更多的宇航员和太空飞船到访过月球。

Since the first moonwalk, other astronauts and spacecraft have visited the Moon.

2009年，"月球环形山观测和遥感卫星"（简称LCROSS）探测器将火箭撞向了月球上的一个深坑之中。

In 2009, a space **probe** called LCROSS blasted a rocket into a deep crater on the Moon.

这次撞击使一大团尘埃和其他物质扬起到太空中。

The crash made a huge cloud of dust and other material rise up into space.

LCROSS研究了这团尘埃，并发现在岩尘中夹杂着细小的冰屑。

LCROSS studied the cloud and discovered tiny pieces of ice mixed in with the rocky dust.

这意味着月球上的陨石坑里有冰冻的水。

This means there is frozen water inside craters on the Moon.

如果月球上有水，人类也许有朝一日能在那里生活！

If there is water on the Moon, people might one day be able to live there!

许多人梦想在月球上建造一座基地或城市。生活在基地里的人们将免受高温和寒冷影响。基地里还有氧气供人们呼吸。

Many people have dreamed of building a Moon base, or city, on the Moon. Inside the base's buildings, people would be protected from the heat and cold. There would also be oxygen for people to breathe.

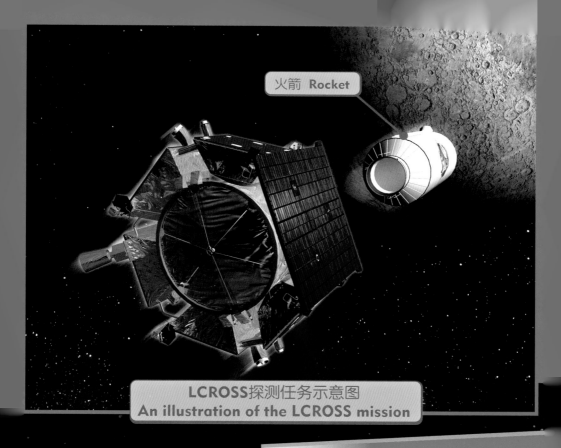

火箭 Rocket

LCROSS探测任务示意图
An illustration of the LCROSS mission

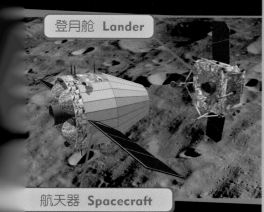

登月舱 Lander

航天器 Spacecraft

2020年12月1日，我国嫦娥五号探测任务将航天器送上了月球。一个小型登月舱造访了月球地表并搜集了月球岩石样本，以供送回地球进行研究。随后航天器于2020年12月17日返回地球。

On December 1, 2020, China's Chang'e 5 mission sent a spacecraft to the Moon. A small lander visited the Moon's surface and collected rocks for scientists back on Earth to study. Then the spacecraft returned to Earth on December 17, 2020.

有趣的月球知识
The Moon Fact File

月球是地球的"近邻"，以下是一些有趣的月球知识。

Here are some key facts about Earth's nearest space neighbor, the Moon.

月球是如何得名的
How the Moon got its name

"月球（Moon）"一词由来已久，原意是"月份"和"度量"。月球绕地球公转一圈的时间长度被用于计量一个月的天数。

The word "moon" comes from very old words meaning "month" and "measure". The Moon's orbit around Earth has been used to measure the number of days in a month.

登月宇航员
Astronauts on the Moon

已有12名宇航员登上过月球。

Twelve astronauts have walked on the Moon.

太阳系里的卫星
The solar system's moons

太阳系里至少有300颗卫星围绕行星运行。这张图显示了五大卫星与月球的大小对比。

There are at least 300 moons orbiting planets in the solar system. This picture shows the sizes of the five biggest moons compared to Earth's Moon.

地球卫星（月球）	木星卫星（盖尼米得）	土星卫星（泰坦）	木星卫星（卡里斯托）	木星卫星（艾奥）	海王星卫星（特里顿）

The Moon Earth	Ganymede Jupiter	Titan Saturn	Callisto Jupiter	Io Jupiter	Triton Neptune

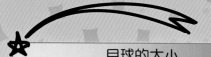

月球的大小
The Moon's size

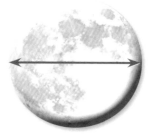

月球的直径约3 475千米

3,475 km across

月球与地球的距离
The Moon's distance from Earth

月球与地球的最短距离是363 104千米。

月球与地球的最远距离是405 696千米。

The closest the Moon gets to Earth is 363,104 km.

The farthest the Moon gets from Earth is 405,696 km.

月球绕地球的轨道长度
Length of the Moon's orbit around Earth

2 413 402.16千米

2,413,402.16 km

月球 The Moon

地球 Earth

月球轨道
The Moon's orbit

月球上的温度
Temperature on the Moon

最高温度：123摄氏度

最低温度：零下233摄氏度

Highest: 123°C

Lowest: -233°C

月球绕地球公转的平均速度
Average speed at which the Moon orbits the Earth

每小时3 681千米

3,681 km/h

动动手吧：制作陨石坑！
Get Crafty : Make Moon Craters!

太空岩石撞击月球表面会形成陨石坑。使用鹅卵石和熟石膏，你也可以制作月球陨石坑。

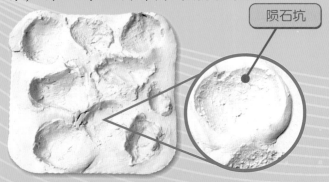

陨石坑

你需要：
- 一只碗
- 熟石膏
- 水
- 一个勺子
- 一个锡纸烤盘
- 报纸
- 鹅卵石或小石头

1. 将熟石膏倒进碗里。加入一些水并搅拌，直至这些混合物看起来像细腻的煎饼面糊。继续添加石膏或水，以取得更多的混合物，因为你需要足够的混合物来装满烤盘。

石膏混合物

2. 将混合物倒入烤盘里，再将烤盘放在地板上（烤盘下垫些报纸），等待混合物成型。不时用手指触碰混合物，当你感觉混合物变得像刚融化的冰淇淋时，就可以动手制造陨石坑了。

3. 将鹅卵石砸进烤盘里。它会立即将石膏混合物砸出了一个坑！小心、迅速地取出鹅卵石，接着换块鹅卵石再砸一次。

4. 当石膏混合物布满坑的时候，将它静置，等它慢慢变硬。这大约需要30分钟。然后小心地将模型从烤盘中取出。

鹅卵石

词汇表 Glossary

小行星 | asteroid

围绕太阳公转的大块岩石，有些小得像辆汽车，有些大得像座山。

宇航员 | astronaut

受过特殊训练，乘坐宇宙飞船进入太空的人。

大气层 | atmosphere

行星、卫星或恒星周围的一层气体。

彗星 | comet

由冰、岩石和尘埃组成的天体，围绕太阳公转。

陨石坑 | crater

圆形坑洞，通常由小行星和其他大型岩石天体撞击行星或卫星表面而形成。

矮行星 | dwarf planet

围绕太阳运行的圆形或近圆形天体，比八大行星小得多。

气体 | gas

无固定形状或大小的物质，如氧气或氦气。

撞击盆地 | impact basin

非常大的陨石坑，直径可达数百千米。

卫星 | moon

围绕行星运行的天体。通常由岩石或岩石和冰构成。直径从几千米到几百千米不等。地球有一个卫星，名为"月球"。

公转 | orbit

围绕另一个天体运行。

氧气 | oxygen

空气中一种无形的气体，是人类和其他动物呼吸所必需的。

平原 | plain

一大片平坦的地面。

行星 | planet

围绕太阳公转的大型天体：一些行星，如地球，主要是由岩石组成的；其他的行星，如木星，主要是由气体和液体组成的。

探测器 | probe

不载人太空飞船。通常被送往行星或其他天体，用于拍摄照片并收集信息，由地球上的科学家操作控制。